# At Home
# on the Farm

## Sharon Gordon

 **Marshall Cavendish**
Benchmark
New York

This is our farm.

It is a busy place.

We work hard all day long.

We start the day with a big breakfast.

Then we feed the animals.

They eat *grain*.

The barn is their home.

We clean it every day.

We collect eggs from the chickens.

Machines pump milk from the cows.

Our sheep have thick, soft hair called *wool*.

We cut it and sell it.

We grow *crops* on our farm.

We plant corn and wheat in the spring.

We water the plants.

They grow tall in the summer.

We *harvest* the crops in the fall.

We keep them in the barn.

We have pets on the farm.

Our horses are fun to ride.

Our dog herds the sheep.

Our cat chases the mice.

We go to the state fair
in the summer.

The best animal wins
a prize.

In the fall, visitors come to our farm.

They always leave with a smile!

# Farm Home

**corn**

**grain**

**harvest**

**horses**

**sheep**

**wool**

# Challenge Words

**crops** (KROPS) Plants that are grown for food.

**grain** (GRANE) The seeds from wheat, corn, rice, or oats.

**harvest** (HAR-vist) To gather crops.

**wool** (WUHL) The thick, soft, curly hair of sheep that can be made into yarn or cloth.

29

# Index

Page numbers in **boldface** are illustrations.

animals, 6–8, **7**, 10, 20–22, **21**, **23**, 24, **25**.
　See also sheep

barn, 8, **11**, **15**, 18
breakfast, 4, **5**

cat, 22
chickens, 10
cleaning, 8, **9**
corn, 14, **15**, **28**
cows, 10
crops, 14–18, **15**, **17**, **19**

dog, 22, **23**

eggs, 10, **11**

fair, 24, **25**
fall, 18, **19**, 26, **27**
farm, 2, **3**

grain, 6, **7**, **28**

harvest, 18, **19**, **28**
herding, 22, **23**
horses, 20, **21**, **28**

machines, 10, **19**
mice, 22
milk, 10

pets, 20–24, **21**, **22**, **25**
planting, 14
prize, 24, **25**

sheep, 12, **13**, 22, **23**, **29**
spring, 14
summer, 16, 24

visitors, 26, **27**

watering, 16, 17
wheat, 14
wool, 12, **13**, **29**
work, 4–12, **7**, **9**, **11**, **13**

## About the Author

Sharon Gordon has written many books for young children. She has always worked as an editor. Sharon and her husband Bruce have three children, Douglas, Katie, and Laura, and one spoiled pooch, Samantha. They live in Midland Park, New Jersey.

With thanks to Nanci Vargus, Ed.D. and
Beth Walker Gambro, reading consultants

Marshall Cavendish Benchmark
Marshall Cavendish
99 White Plains Road
Tarrytown, New York 10591-9001
www.marshallcavendish.us

Library of Congress Cataloging-in-Publication Data

Gordon, Sharon.
At home on the farm / by Sharon Gordon.
p. cm. — (Bookworms. At home)
Includes index.
Summary: "Describes life on a farm, including grain, wool, and crops"—Provided by publisher.
ISBN 0-7614-1958-6
1. Farm life — Juvenile literature. 2. Agriculture — Juvenile literature.
I. Title II. Series: Gordon, Sharon. Bookworms. At home.
S519.G634 2005
630—dc22
2004025382


Photo Research by Anne Burns Images

Cover Photo by *Corbis*

The photographs in this book are used with permission and through the courtesy of:
*Corbis*: pp. 1, 15, 28 (upper l.) Richard T. Nowitz; pp. 5, 21, 28 (lower r.) Royalty Free;
pp. 7, 28 (upper r.) MacDuff Everton; p. 11 Richard Hamilton Smith; pp. 13, 29 (right) Amos
Nachoum; p. 17 Craig Aurness; pp. 19, 28 (lower l.) Philip Gould; pp. 23, 29 (left) Kevin R. Morris;
p.25 Chris Jones. *Index Stock Imagery*: p. 3 Blue Water Photo; pp. 9, 27 Frank Siteman.


Series design by Becky Terhune

Printed in Malaysia
1  3  5  6  4  2